DOES IT FART?

THE DEFINITIVE FIELD GUIDE TO ANIMAL FLATULENCE

DOES IT FART?

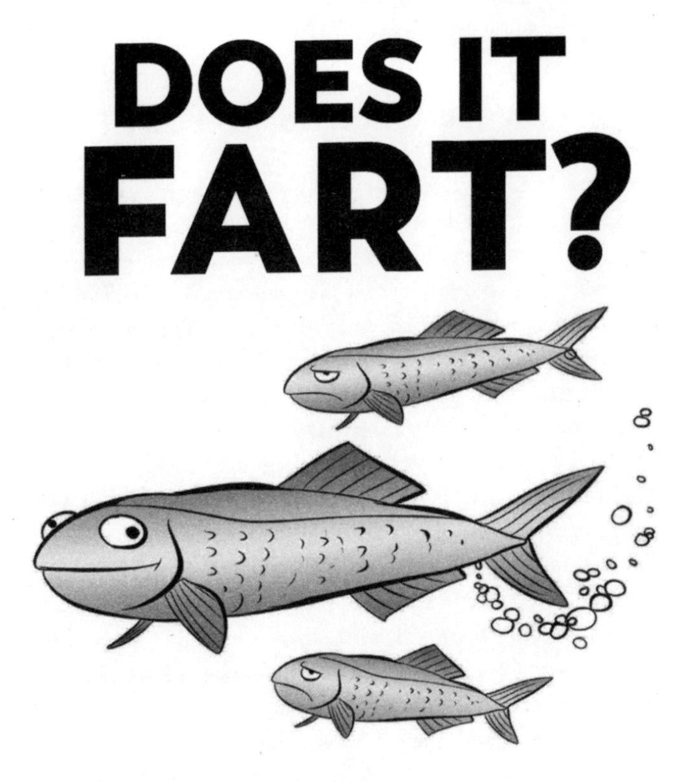

WRITTEN BY NICK CARUSO & DANI RABAIOTTI

ILLUSTRATED BY ETHAN KOCAK

Quercus

INTRODUCTION

HOW THIS BOOK CAME ABOUT

Nick and Dani are both active on Twitter and use the social media platform to talk about their work and engage with other scientists. They are part of the large community of ecologists and zoologists on Twitter who share information, collaborate and engage in a lot of science communication. One fateful day Dani was asked by a family member whether snakes farted, but she was unsure of the answer. She knew someone, however, who would definitely know: David Steen, an Assistant Research Professor of Wildlife Ecology and Conservation at Auburn University, Alabama, and all-round snake expert. He answered, on Twitter, "<sigh> yes", and from there, science Twitter quickly realised that actually 'does it fart?' is a common question that animal researchers get asked: this in turn led Nick to create #DoesItFart and, in the true nature of science, this swiftly spawned a spreadsheet. Many science researchers and pet owners contributed to the spreadsheet: You can find a list of contributors and their Twitter handles on page 132–33. Clearly the next step was a full guide in print, and so *Does it Fart?*, the book, was born.

WHAT IS A FART?

The medical term for a fart is 'flatulence', which is defined as 'flatus expelled through the anus'. Flatus is strictly defined as gas produced during digestion – generally in the stomach and/or gut. Through this book Dani and Nick are taking part in flatology, or the study of flatulence, even though their main area of expertise is not in this field.

The word 'fart' dates back to the fourteenth century, before the term flatulence came into use, and was used specifically to mean breaking wind loudly. Today the term fart is more commonly used to describe any gas expelled from the end of an animal that is opposite to its mouth – whether this be through the anus, cloaca or a specialised duct – and if it is audible or not. This, therefore, is the definition used throughout this book. So, although some of the farts in this book may not fit the strict medical definition of flatulence, they would generally be accepted as a fart to any human encountering them.

Not all farts are created equal, and the smell and frequency of flatus can vary based on an organism's diet, health and gut flora. Vegetables and other food that are high in fibre, such as broccoli, beans or peas, dairy products containing lactose, foods that are high in starch or fructose and many others, have all been linked to increasing fart frequency in humans (and likely other animals too, though studies are limited). We all know the childhood song 'Beans, beans, good for your heart, the more you eat the more you fart', similarly, food that is harder to digest for a particular organism and remains within their intestines for longer periods of time can also lead to increases in farting. While many farts are odourless, consisting primarily of carbon dioxide, food containing a higher concentration of sulphur, such as meats or Brussels sprouts, can lead to the production of hydrogen sulfide, which has a pungent rotten-egg smell. Parasitic infections, such as Giardiasis, other intestinal illnesses and food sensitivities can also lead to increases in flatus frequency or malodour – that is, having a bad smell. Whereas some individuals may simply harbour a greater concentration of gas-producing bacteria and other microorganisms within their intestines, collectively known as the intestinal or gut flora, and thus produce more gas.

CONTENTS

HERRING

Scientific name (Genus): Clupea

DOES IT FART? YES

There are over 200 species of herring – fish species in the genus *Clupea* – worldwide, a couple of which have taken the art of farting to new depths. The Pacific and Atlantic herrings have both been found to gulp air from the surface of the water and store it in their swim bladders, later expelling it from their anal ducts in what is technically known as a Fast Repetitive Tick (FRT).

FRTs produce a 'high-pitched raspberry sound' lasting between 0.6 and 7.6 seconds, at frequencies of between 1.7 and 22kHz. Herring, which have particularly good hearing compared to other fish species, produce FRTs more frequently when densities of herring are higher, and it is thought that they use FRTs to communicate with other members of their species. In this way, FRTs are used to find and stay close to other herring, especially at night when they can't see each other; so these particularly gassy fish use farts to form shoals and stay safe from predators. You might think that these raspberry sounds would give away a herring's location to hungry predators looking for a bite to eat, but, due to the high frequencies of the FRTs, they are above the auditory range of most predatory fishes – a secret fart-code that only other herring can hear. Most marine mammals (as well as humans!) can hear it though, and it is thought this might be how they locate herring to eat.

GOAT

Scientific name (Species): Capra aegagrus hircus

DOES IT FART? YES

Goats belong to the family *Bovidae*, which includes cows (page 102), meaning they have four stomachs packed full of methane-producing bacteria which help them digest plant material, and so give off a lot of gas in the process. Although this process produces far more burps than farts, goats do fart, and this combination makes them particularly gassy animals. In 2015, a plane full of over 2000 goats on its way to Kuala Lumpur was forced to land unexpectedly after the fire alarm was set off by the copious amounts of gas produced by the goats on board.

Domestic goats, and their farts, have lived alongside humans for over 10,000 years, thanks to the goat's hardy nature and milk production. One of the oldest surviving secular songs in the English language – 'Sumer is icumen in', which is about the sights and sounds of summer – includes the line 'Bulluc sterteþ, bucke uerteþ', which is thought to mean 'Bullock leaps, billy-goat farts'. So it isn't just goats that are culturally important, but their farts too.

SONORAN CORAL SNAKE

Scientific name (Species): Micruroides euryxanthus

DOES IT FART? YES

This brightly coloured, highly venomous snake species is relatively widely distributed and can be found across southern Arizona, parts of New Mexico, Sonora and the surrounding areas in Mexico. As with most venomous snakes, this species isn't actually very keen on biting, instead it uses a highly unusual behaviour as its first line of defence against predators. If threatened, the snake hides its head under its body, raises its tail and sucks air into its cloaca (the part of the body from which snakes defecate and urinate) before forcefully expelling it again. This emits a popping sound of around 2.5kHz, known as 'cloacal popping'. These pops sound like a higher-pitched, shorter version of a human fart (see page 118) and can be heard from up to 2 metres away! Sadly, the effectiveness of this behaviour at keeping predators at bay has never been measured, but between that, the bright colours and the venom, they seem to do fine.

Cloacal popping is relatively rare among snakes but it has also been observed in the Western hooknose snake, *Gyalopion canum*, which, when startled, simultaneously thrashes around and defecates whilst emitting cloacal pops. So really the Sonoran coral snake is positively polite in comparison…

BABOON

Scientific name (Genus): Papio

DOES IT FART? YES

The genus *Papio*, commonly known as baboons, is made up of five species found throughout Africa, as well as, in the case of the Hamadryas baboon (*Papio hamadryas*), parts of the Arabian peninsula. Baboons have been on the planet for at least two million years and are highly social, living in groups, known as troops, of up to 250 individuals (although generally group sizes are closer to 50 or so) with complex dominance dynamics regulating group life – dynamics in which farting can play an unexpectedly significant role. As with most primates, baboons fart often and unashamedly. When female baboons are in heat their sexual organs and rump swell up, indicating to males that they are ready to mate, and this reportedly makes any flatulence more potent, or potentially just more audible. And people say romance is dead!

Within a troop there are frequent fights between males and during a disagreement subordinate males will often flee from a dominant male, screaming, defecating and farting as they run. As with chimpanzees (page 63), researchers have been known to locate troops by listening out for their audible flatulence, something that is useful given that baboons are surprisingly good at disappearing into any nearby vegetation.

MILLIPEDE

Scientific name (Class): Diplopoda

DOES IT FART? YES

Millipedes get their scientific name *Diplopoda* from the fact that, unlike other arthropods, they have two pairs of legs on each segment of their body. Also unlike many other arthropod groups, millipedes have a very simple digestive system which lacks a pouch in their hind-gut for food storage. This means food passes through millipedes quickly and so must be broken down as fast as possible. To help, millipede intestines contain a type of organism called methanogenic archaea – single-celled microbes which assist in breaking down their food (mostly rotting wood and leaf litter), producing methane in the process.

Different species of millipede have different types of archaea in their gut, and methane production is positively correlated with body mass – that is, the bigger the millipede, the bigger the farts. As with many other groups of insects, millipedes living in the tropics are bigger than those living in temperate climates, therefore tropical species tend to produce more gas. The biggest species of millipede – the Giant African millipede, *Archispirostreptus gigas* – can grow up to 38 centimetres long, with about 256 legs, and lives (and, we assume, farts) mainly in the lowland forests of East Africa.

BEADED LACEWING

Scientific name (Species): Lomamyia latipennis

DOES IT FART? YES

There are hundreds of species of the little winged insects called beaded lacewings, and very little is known about their biology, especially at the larval stage. What is known about this incredibly understudied taxon, however, is that it is found on every continent except Antarctica. If that distribution sounds familiar, it is because it is shared by another insect species, the termite (page 37). A number of beaded lacewing species have been found to have a larval stage which live in close association with termites. The adult beaded lacewing lays her eggs on rotting wood next to a termite nest and the larvae hatch out and proceed to sneak into, and live inside, the nest, preying on the workers in a variety of gruesome ways.

One species, *Lomamyia latipennis*, has a particularly ingenious way of stunning and killing its prey: it farts on them. The larva raises its tail towards the termite's head and releases a potent allomone (a chemical that affects termite behaviour) which paralyses the termite and, ultimately, kills it. This allomone does not affect any other species of insect, or the larva itself, so this species of lacewing has evolved to produce a very specific chemical fart perfectly designed for its larval life inside the nest of its prey, one of the very few genuinely fatal farts known to science.

HORSE

Scientific name (Species): Equus ferus caballus

DOES IT FART? YES

Horses are among the most frequent farters in the animal kingdom. This is because, unlike bovids – the family which includes cows (page 102), antelope and llamas (page 58) , among others – horses are what are known as hind-gut fermenters. This means that the plant matter they eat passes through the stomach and intestines before being digested in the hind-gut through fermentation. Plants are particularly hard to digest due to the fact they contain a high concentration of cellulose, which means that they must be broken down in the horse's colon by a wide variety of gut flora (bacteria and archaea).

One big by-product of any fermentation process is gas – and horses produce a lot of it! They have especially long colons, around 3.5 metres long, to help them digest all that plant material. This gives extra time and space for gas build-up, and because the colon is at the end of the digestive tract this gets released copiously, and frequently, in the form of farts. Horses will fart any time, any place, as anyone who has spent time with them will know. One positive side-effect of these gut bacteria is that they also provide horses with all the vitamins and minerals they need to stay healthy, meaning the farts are only a minor inconvenience.

KANGAROO

Scientific name (Genus): Macropus

DOES IT FART? YES

Kangaroos once had a key role in fart science. For a number of years researchers were diligently attempting to lower cow methane emissions by transplanting gut bacteria from kangaroos into cows. Why? Because it was believed that kangaroo farts contained very low levels of methane, and cows emit so much methane that they are a significant contributor to climate change. However, more recently a study by Dr Adam Munn, at the University of Wollongong, showed that kangaroos actually produce more methane through their flatulence than was previously thought. Although this is still less than cows (page 102) and other ruminants, it seems that, per kilogram of animal, this is equivalent to a number of other species, including hind-gut fermenters such as horses (see opposite). Furthermore, this lower methane production is more likely a product of kangaroo digestive anatomy, which uses fore-gut fermentation, where plant matter is broken down anaerobically (without oxygen) by bacteria in the fore-stomach before reaching the rest of the digestive system. This means that food passes through the kangaroo very quickly, leaving less time for gas to be produced. So, sadly, it seems kangaroo farts are not the answer to the issue of climate change after all!

BOLSON PUPFISH

Scientific name (Species): Cyprinodon atrorus

DOES IT FART? YES

Cyprinodon atrorus is a species of pupfish that is endemic to – that is, it only lives (and farts) in – small shallow pools in the Cuatro Ciénegas Reserve in Northern Mexico. It turns out that these fish have one of the most appropriate names in the animal kingdom, because the word 'pups' is German slang for fart. C. *atrorus* feed on algae and other organisms found in the sediment. The pools have constantly changing temperatures and levels of salinity. In summer, when temperatures are at their highest, the algae produce gas bubbles which the pupfish ingest with their food. This results in a build-up of gas in their bodies, inflating their intestines and distending their abdomen, causing them to lose equilibrium and have trouble swimming, so they begin to float. These fish like to bury themselves in the sediment, but when they are full of gas they repeatedly pop back up, out of the sand and float to the surface. The only relief comes through farting, at which point the fish is able to right itself and swim normally. If the fish are unable to fart out the gas, their inability to stay away from the water's surface puts them at risk of predation by birds such as herons and high levels of trapped wind can cause death through intestinal rupturing. Up to 300 fish have been found to have died simultaneously in this way, so although this species does fart, they would benefit from doing so more often, as it really is a case of fart or die.

AFRICAN WILD DOG

Scientific name (Species): Lycaon pictus

DOES IT FART? YES

The African wild dog is a highly social canid species that lives in groups of between 2 and 26 individuals. They are cooperative breeders; a single, dominant pair mates and produces pups while the rest of the pack help to raise the litter. Wild dogs often hunt as a pack, with the whole group helping to bring down prey much bigger than themselves, including impala and wildebeest. When the pack has pups in the den they leave behind a babysitter to guard them against predators, while the rest of the pack brings back, and subsequently regurgitates, food for both the babysitter and pups.

Upon returning from the hunt the whole pack greet each other and become very excited (who wouldn't get excited at the prospect of regurgitated meat for dinner?). One side-effect of this is that they defecate everywhere, often accompanied by bouts of flatulence. One scientific paper from the 1950s goes as far as to note 'an objectionable smell renders them somewhat unpleasant as household pets' (though this is far from the only reason why this would be a bad idea). Wild dogs smell particularly strongly even without the additional farts, but the jury is out among wild-dog researchers as to whether this smell is pleasant or unpleasant.

BAT

Scientific name (Order): Chiroptera

DOES IT FART? MAYBE

Over 1200 species of bat have been discovered globally to date, but there are likely to be far more in existence, as some species are incredibly difficult to tell apart. Researchers often have to look at their tiny teeth, or in some cases record their echolocation calls, to distinguish between species. You may sometimes hear apparent farting sounds from bats, however these raspberry-like noises are actually coming from the front end. The echolocation calls of *Microchiroptera* (small bats: most *Megachiroptera* or fruit bats, do not echolocate) consist of a wide variety of sounds, including those resembling a high-pitched bout of flatulence.

As mammals, it seems likely that bats do fart, and they certainly have the right bacteria present in their gut. However, bat digestion is incredibly quick; they have a very short digestive system because carrying around a lot of extra food weight when flying uses up a lot of energy. Even in the largest bat species, the flying fox (genus *Pteropus)*, which can weight up to 1 kilogram, digestion only takes 12–34 minutes from mouth to anus. This could mean that bats don't fart, or if they do, it may not be in audible quantities – certainly there appears to be no positive confirmation of bat farts in the scientific literature. What is clear, however, is that if bats do fart, the bigger the bat, the bigger the toots.

PORTUGUESE MAN O' WAR

Scientific name (Species): Physalia physalis

DOES IT FART? NO

Portuguese man o' war may look similar to jellyfish, but although they are in the same phylum, Portuguese man o' war are not, in fact, jellyfish, or even in many senses an animal at all. Despite appearing to be a single organism they are actually made up of a colony of lots of tiny specialised organisms known as zooids. Portuguese man o' war catch their food, mostly small fish, using their stinging tentacles which are made up of specialised stinging dactylozooids. The food is then lifted to a different type of zooid – the gastrozooid – which secretes digestive enzymes all over the food, slowly liquefying it (tasty!).

During this process there isn't any opportunity for the production or build-up of gas, and because man o' war (or man o' wars – both are correct) lack an anus and a digestive system in general, they certainly cannot fart. That said, they do have one very gassy feature: a gas-filled bladder known as a pneumatophore, which stays floating on the surface of the sea, propelling the man o' war along wherever the wind blows.

PARROT

Scientific name (Order): Psittaciformes

DOES IT FART? NO

Here's where the science starts to get less clear-cut than you might like. As you will learn on page 57, birds do not fart. When collecting the data that went into this publication, however, there were numerous reports of farting parrots, and numerous examples of this activity can be found on the internet. So what is going on?

Well, parrots are exceptionally good at mimicking sounds made by humans, as well as other animals, and even noises such as the television. One African grey parrot (*Psittacus erithacus*), Alex, learnt over 100 words and was thought to be able to identify objects and colours. Prosecutors in Michigan in 2016 were even considering using a parrot as evidence in a murder trial when he started repeating the words 'don't shoot' after his owner was shot and killed. So the reported cases of parrot farts are more than likely 'hot air'; instead, parrots are mimicking the sound of humans farting – meaning any farts you hear from a parrot are coming from their mouth, not their cloaca!

UNICORN

Scientific name (Mythical): Monocerus

DOES IT FART? YES

It would seem sensible to assume that since horses (page 10) fart, unicorns, often described as horses with a single horn emerging from the centre of their head, must also fart.

The origins of the unicorn date back not to Greek mythology but actually to Greek natural history, where scholars described them as living in the forests of India. Today, it seems likely that these reports were in fact based on sightings of the Arabian oryx, *Oryx leucoryx*; potentially of individuals that had lost one of their horns whilst fighting. Oryx are part of the family *Bovidae* (the same family as cows, page 102), and do fart, so again it is likely that the same would apply to unicorns. On the other hand, the unicorn myth may have emerged from descriptions passed down through generations of a large ice-age rhino species, *Elasmotherium*, which had a single, large horn in the centre of its head. As extant rhinos (page 24) fart copiously, we can assume that this extinct species did too. Unicorns may not exist, but if they did they would definitely fart. Scientists are still debating whether those farts would be made up of rainbows and glitter.

SEA ANEMONE

Scientific name (Order): Actiniaria

DOES IT FART? NO

Anemones don't have an anus, or much of a digestive system, so strictly speaking they do not fart. They have one opening into their gastrovascular cavity – the anemone equivalent of the stomach – where food is digested. This opening, the siphonophore, does double duty as both a mouth and an anus. Hapless small animals are trapped in the anemone's tentacles, which contain stinging cells called nematocysts, and are dragged into the gastrovascular cavity, where they are digested.

The bad news, if you are sea anemone prey at least, is that the stinging isn't over, because this cavity contains stinging filaments called acontia which also release enzymes to break down food. In this way anemones can digest their food in as little as 15 minutes, and any food the anemone can't digest, such as shells or bones, is passed back out through the mouth (or anus, depending on how you want to look at it). If the anemone is threatened it will eject its acontia filaments from its siphonophore as a form of defence, stinging and repelling any would-be predators. Sadly, there is no gas involved, or that really would be a silent but deadly fart.

SPIDER

Scientific name (Order): Araneae

DOES IT FART? NOBODY KNOWS

Spider flatulence is an oddly understudied topic in scientific literature, but we can look to their digestive system for some clues. Spiders do the vast majority of digestion outside their body, injecting their prey with venom from their fangs before ejecting sputum, full of digestive enzymes, from their mouths, through the fang holes and into the body of their prey. They then wait whilst the digestive juices break down tissues inside the exoskeleton or, in some cases, skin of their prey. Spiders will then suck up the liquidy goodness into their mouths and stomachs, then they regurgitate it and eat it again. This happens a number of times, as spider digestive systems can only handle liquids – which means no lumps! It would seem likely that spiders ingest air during this process – one of the key elements needed for a fart.

Once they have extracted all the nutrients, in an organ called the ceaca, the food passes to the stercoral sac, where the moisture is extracted before anything left over is excreted through the anus as waste. Since the stercoral sac contains bacteria, which helps break down the spider's food, it seems likely that gas is produced during this process, and therefore there is certainly the possibility that spiders do fart. No work has been done to verify this to date, however, so the truth remains a mystery until urgently needed research funding is allocated.

RHINOCEROS

Scientific name (Family): Rhinocerotoidea

DOES IT FART? YES

There are five species of rhino alive today: the more well-known black (*Diceros bicornis*) and white (*Ceratotherium simum*) rhinoceroses of Africa, but also the Indian (*Rhinoceros unicornis*), Javan (*Rhinoceros sondaicus*) and Sumatran (*Dicerorhinus sumatrensis*) rhinoceroses of Asia. Similar to horses (page 10), rhinos are hind-gut fermenters, meaning they browse constantly on plant material, breaking it down at the end of their gut, after it has passed through the stomach and small intestines. As rhinos are much bigger than horses, they produce a lot more gas. When in the African bush in close proximity to rhinos it is not uncommon to hear them emit numerous loud, long farts while they graze. Rhino farts also smell really bad, so much so that they have even given rise to a piece of brewing terminology; when the yeast used to make alcohol through fermentation produces hydrogen sulphide it gives off a horrible sulphur smell, known as a rhino fart.

Sadly, there are far fewer rhino farts in the world today than there were historically, as large numbers of all five species have been killed for their horns. In fact, so much so that the Sumatran, Javan and black rhinos are all critically endangered, with less than 250 individuals of the Sumatran and Javan rhinos in particular left in the wild.

ELEPHANT

Scientific name (Family): Elephantidae

DOES IT FART? YES

There are two species of elephant: the African (*Loxodonta africana*) and Asian (*Elephas maximus*). In parts of Asia elephants are often domesticated and used for work such as carrying heavy logs. As you can imagine for an animal of such a large size, elephants produce incredibly pungent farts in great volume. This being the case, people who work with elephants – known as mahouts – have learned to treat particularly potent elephant flatulence by feeding them rice mixed with grilled garlic, although how and why this works is unknown.

In the wild, elephants need to spend most of their time browsing on low-quality vegetation which is high in cellulose and difficult to digest. Elephants are hind-gut fermenters, like rhinos (page 24) and horses (page 10), and as a result have a very long digestive tract, full of bacteria that help them break down tough-to-eat foods such as tree bark. This is part of the reason why elephants were able to become so large; their digestive physiology means that instead of being held in the stomach, as it is in ruminants, food passes straight through to the colon and therefore through their body much faster. Digestive systems that use hind-gut fermentation can accommodate much larger food intakes, enabling animals to evolve larger body sizes, so it is thanks to this incredibly gassy digestive system that these amazing animals are around today.

BEARDED DRAGON

Scientific name (Genus): Pogona

DOES IT FART? YES

Bearded dragons make up the genus *Pogona*, a group of lizards that hail from central Australia. They are highly territorial, warding off competitors by puffing out their 'beard' (a flap of skin under their chin which they inflate to show aggression) which is found in both sexes. They are popular pet species, most commonly the Central bearded dragon (*Pogona vitticeps*).

Bearded dragons sometimes fart audibly, usually accompanied by defecation, and this behaviour is particularly noticeable when they are submerged in water. Many pet owners have reported that vivariums (a tank where pet reptiles are kept) smell particularly bad after a bout of bearded-dragon flatulence.

In the wild these animals have an incredibly varied diet, consuming small lizards, insects, mammals, fruit and flowers. They have a mild venom which they use to subdue larger prey, however this is not harmful to humans. In captivity they are generally fed on a variety of fruit and vegetables, supplemented with insects, and it has been reported by some owners that butternut squash is a recipe for particularly pungent farts.

CHEETAH

Scientific name (Species): Acinonyx jubatus

DOES IT FART? YES

Cheetahs are famed as the fastest animal on land. It is a species of spotted cat historically found throughout most of Africa, as well as across the Arabian peninsula and into India. Today, sadly, the cheetah has been restricted to just 10 per cent of its former range in Africa and the central deserts of Iran, with just an estimated 6700 cheetahs left in the wild.

Cheetah, as with other cats (see cat, page 85), lions (page 34), bobcats (page 82) and snow leopards (page 101), eat a diet that consists entirely of meat – cheetah prey mostly on antelope species such as gazelle and impala. Digestion of high volumes of meat produces high levels of putrefaction compounds, leading to some particularly pungent farts. One study into cheetah digestion found that the inclusion of non-digestible animal parts, such as cartilage, bone and collagen, was particularly important in aiding in the fermentation of food in the cheetahs' gut. As a result greater quantities of gas are produced through giving the gut flora a surface to attach to, enabling increased fermentation – and therefore likely increasing fart production.

ZEBRA

Scientific name (Genus): Equus

DOES IT FART? YES

There are three species of zebra alive today: plains (*Equus quagga*), Grevy's (*Equus grevyi*) and mountain (*Equus zebra*). Zebras are probably best known for their distinctive black-and-white-striped coat, and there have been a number of theories proposed by scientists as to why they evolved this distinctive colouration. One theory is that the stripes could either camouflage them in dappled shade or, more likely, be used to confuse predators when the animals are running. Within a zebra species individuals can recognise other individuals from the configuration of their stripes, so these patterns are likely used to some extent for identification. More recently, these stripes have also been found to deter biting flies.

It is unclear which is the primary driver in stripe evolution, but it seems that zebras' black-and-white colouration is advantageous in a number of ways. As you may have noted, zebra belong to the same genus as the domestic horse (page 10), *Equus*, and both animals share similar farting habits. Zebra farts can be heard from long distances across the plains of Africa, and this is particularly common when they are startled and begin to run – the motion propels the gas from their bodies and they often fart loudly with each stride.

DINOSAUR

Scientific name: (Clade) Dinosauria

DOES IT FART? NOT ANY MORE

The dinosaurs were an incredibly diverse clade of reptiles that walked the Earth from between 231 and 243 million years ago, until a mass extinction event wiped out most species. Birds are the direct descendants of feathered dinosaurs, and as we know birds do not fart (page 57), so we might assume that dinosaurs, or at least those species from which birds descended – the Maniraptoran dinosaurs – did not fart. However, another group of dinosaurs, the sauropods, almost certainly passed gas. Like today's large herbivores, they fed exclusively on plants, and their large size suggests that they had a similar digestive anatomy – using hind-gut fermentation to break down the cellulose in their food. It seems likely that these sauropods hosted methanogenic bacteria in their gut in order for them to be able to obtain enough energy from their food. One study suggests that they may have farted out as much as 1.9kg of methane per dinosaur per day!

It is difficult to determine just what bacteria were present in these animals' guts though, due to the fact they died out over 66 million years ago. But one thing is certain, they definitely don't fart any more.

LION

Scientific name (Species): Panthera leo

DOES IT FART? YES

Lions (*Panthera leo*) are known as the king of the jungle, although, contrary to what this title would suggest, lions actually don't generally live in the jungle but mostly inhabit savanna, scrub and dry woodlands in Africa and India. On top of this, it is actually female lions that do the vast majority of the hunting, not the males, with some studies suggesting females make up to 90 per cent of the kills within a lion pride. Meanwhile the males sleep up to 20 hours a day! Lions, like other cats, are true carnivores; that is, they eat nothing but meat. This is reportedly a recipe for particularly strong-smelling farts. In the wild lions live to 10–14 years old, with the females living longer than the males. In captivity lions can reach 30 years old, and reportedly their bouts of flatulence become increasingly common with age.

Another particularly smelly behaviour that lions engage in is scent marking, where a male lion sprays urine and rubs faeces around his territory, to mark it as his own. Male lions can spray objects up to 3 metres away – another good reason to keep your distance from lions, if the large claws and teeth weren't enough!

GOLDFISH

Scientific name (Species): Carassius auratus

DOES IT FART? NO

Goldfish have become incredibly popular pets since their domestication nearly 1000 years ago, and over 30 million of them are kept as pets in the UK alone. In ancient China it was common to keep carp for food, and occasionally these carp, through a chance mutation, were born a yellow or orange colour. The first record of these 'gold' fish dates back to the year AD 975. Golden fish were seen as good luck, and by the year 1240 people were actively breeding them for ornamental purposes. Today over 300 breeds of goldfish exist, and it is the most popular fish species to own as a pet!

Despite the high number of goldfish living in people's homes, observing them pass gas is a rare occurrence. Although these fish do have gas-producing bacteria in their gut, it is far more common to see them burp out gas than pass it through their anal duct. It is thought that the rarity of this behaviour may be caused by an ability to incorporate digestive gas into their faecal pellets, which are encased in a layer of mucus. So if you have a farting fish it may be having digestive problems!

TERMITE

Scientific name (Infraorder): Isoptera

DOES IT FART? YES

Termites fart a lot: or, at least, termites fart and there are a lot of them – the total weight of these insects on Earth today is greater than that of humans. Their farts are a source of atmospheric methane, a gas which contributes to climate change, and each termite produces around half a microgram (that's half of one-millionth of a gram!) of methane every day. This may not seem like a lot, but as some of the most successful organisms on the planet, with colonies on every continent except Antarctica and colony sizes of up to millions of individuals (depending on the species), this quickly adds up. They are only thought to produce around 5 to 19 per cent of methane emissions globally (around 0.27 per cent of greenhouse gas emissions), but that's still a pretty impressive contribution from such a tiny animal! Given termites evolved over 100 million years ago, they've contributed a lot of methane to the atmosphere since they came into existence. Then again, since anthropogenic emissions through agriculture, burning fossil fuels and waste disposal are shown to make up 63 per cent of current annual methane emissions, we can't really blame the termites for rising global temperatures: that's on us.

Still, termites don't have it all their own way when it comes to farting; one of their predators, a species of beaded lacewing larva (see page 8), kills its victims with toxic, immobilising farts.

WHALE

Scientific name (Infraorder): Cetacea

DOES IT FART? YES

As you can imagine, whale farts are incredibly large. Blue whales, *Balaenoptera musculus*, which are currently the largest animal on the planet, likely emit the largest volume (per fart) of any extant species. Whales, in accordance with their body size, have huge digestive systems, containing a number of stomach chambers which, in the case of the blue whale, combined can hold up to one tonne of food (plenty of room for Jonah, although he wouldn't have fitted down the whale's oesophagus). These chambers are packed full of bacteria which break down the food – either plankton in the case of baleen whales, or fish in the case of toothed whales – producing a whole lot of gas in the process.

Given their size, whale farts are remarkably elusive and have only been captured a handful of times on camera. Whale researchers who have been downwind of these bouts of flatulence report them to be incredibly pungent. It could be worse though: farts aren't the only time that whales have caught people off guard with their gas. When whales die they are often washed up on beaches, where they quickly begin to decompose. This causes huge amounts of gas to build up in their bodies, and some dead whales have been known to explode. One incident, in 2004 in Taiwan, saw a dead whale explode whilst being transported through the centre of Tainan City, showering buildings and onlookers with decomposing whale innards.

SS SMELT IT

AFRICAN BUFFALO

Scientific name (Species): Syncerus caffer

DOES IT FART? YES

The African buffalo is among the largest species of ruminants; an adult male can weigh up to 1000 kilograms. African buffalo eat mostly grass, which is hard to digest and must be consumed in huge quantities to provide enough energy to power such a large animal. Similar to cows, they have four stomach chambers, therefore they produce large quantities of potent farts and burps – one study found that they can produce up to 318 litres of gas per day, enough to fill a large fridge-freezer. As buffalo live in herds of up to 1000 individuals, that's a whole lot of farts!

This isn't the most fearsome thing about the African buffalo, though; they have huge fused horns which, combined with their large size, makes them a fearsome adversary for any would-be predator. As a result, adult buffalo are only preyed upon by very few species – namely lions (page 34) and large crocodiles and, very occasionally, particularly fearless groups of hyenas (page 80). Buffalo can be dangerous to humans, too, due to their unpredictable behaviour (especially of large males) and the fact that, for their size, buffalo are surprisingly good at hiding in trees and scrub. Sometimes, though, loud expulsions of gas from either end can alert you that a buffalo is near.

BROWN RAT

Scientific name (Species): Rattus norvegicus

DOES IT FART? YES

The brown rat, *Rattus norvegicus*, is thought to originate in Asia but has been transported by humans all over the world to every continent except Antarctica. Pet rats – known as fancy rats – are a domesticated version of the brown rat. Fancy rats were originally bred for blood sports in the UK in the eighteenth and nineteenth centuries – people would bet on how long it would take for a terrier to kill all the rats in the ring. Eventually people (including Queen Victoria's own rat catcher, Jack Black) started breeding them specifically for interesting coat colours, leading to the pet rats we know and love today.

Despite their popularity with the prim Victorians, rats fart regularly, and this can certainly be smelt by the human nose – especially when the rat is farting on their owner, which they are often inclined to do. These potent bouts of rat flatulence certainly bring a new meaning to the phrase 'I smell a rat'! One study by researchers looking at gas production in lab rats found that rats produced much higher levels of flatulence when fed on beans, just like humans, but on a smaller scale. This is due to the fact that beans contain high levels of oligosaccharides, a type of sugar that rat (and our) digestive systems have difficulty breaking down, producing a lot of gas in the process.

HONEY BADGER

Scientific name (Species): Mellivora capensis

DOES IT FART? YES

The honey badger, also known as the ratel, is known for its fearsome demeanour and intelligence. Honey badgers are one of the most fearless animals around, they have been known to chase away animals as large as a lion (page 34), or even a buffalo (page 40). Honey badgers will eat anything: honey (naturally), amphibians, lizards, berries, birds, eggs, insects, carrion and even venomous snakes.

Honey badgers are well adapted for their risky lifestyle; they have huge digging claws, powerful jaws full of sharp teeth and incredibly thick skin, which means they are incredibly hardy and difficult to injure. One other, lesser-known, tool that they have in their arsenal, however, is incredibly powerful anal scent glands. These glands are used by the honey badger to mark its territory but also to secure its favourite food: honey. The smell from these glands is so potent that it is used by the badger to subdue bees in their nests – after a honey badger attack bees can often be found huddled in a corner of their nest, far away from the pungent smell. It was previously rumoured that the smell was so strong it could kill a whole hive of bees! These reports, however, have since been proven to be untrue. So, although these animals do fart (the smell of which is reportedly pretty strong), that is certainly not the smelliest thing about them!

GIRAFFE

Scientific name (Genus): Giraffa

DOES IT FART? YES

Historically the giraffe was believed to be a single species, but a study into giraffe genetics carried out in 2016 found that there were at least four distinct species, which can generally be distinguished from each other by their unique coat patterns.

Giraffes are the largest of all ruminant species; the largest male ever recorded stood at almost 6 metres high, and they can weigh over 1100 kilograms. As a result of their large size, they have huge stomachs packed with microorganisms which are specially evolved to digest plant matter, producing large quantities of gas in the process. It is unlikely, however, that they produce the largest quantities of gas of any ruminant species, as they are very selective feeders, browsing only on easily digestible plant parts such as fruit and flowers, mostly of *Acacia* tree species. This means that giraffe digestion is faster than in a number of other ruminant species, such as the African buffalo (page 40), and therefore there is less time for gas to be produced. Still, giraffe farts, as with those of most ruminants, have a characteristic strong smell. It is unlikely that this bothers the giraffe, though, as they are probably unable to smell it, given that their nostrils are so far away from their, or any other giraffe's, butts! Evolution is a wonderful thing.

STRIPED SKUNK

Scientific name (Species): Mephitis mephitis

DOES IT FART? YES

Skunks are well known as the smelliest of all animal species. The striped skunk, *Mephitis mephitis*, is one of 12 species in the order *Mephitis* (the skunks) and it lives throughout Canada, the United States and Northern Mexico. Striped skunks feed on a wide variety of food – mostly insects but also small mammals, amphibians, reptiles, eggs, berries, nuts, roots and, in some coastal populations, even fish and crabs. Skunks living in urban areas often supplement their diet with rubbish, which would seem like a recipe for especially stinky farts.

The strong smell that skunks are infamous for, however, is not caused by their flatulence. Skunks have two glands on either side of their anus, filled with a foul-smelling liquid made up of sulphur-containing chemicals known as thiols. Two powerful muscles force this liquid out of the skunk when it is threatened, spraying it distances of up to 3 metres. The scent is so strong it can be smelled by the human nose up to a mile away. This is an incredibly effective predator defence; very few animals will attack and eat a skunk, although some birds of prey, such as the horned owl, are stealthy enough to snatch one and avoid being sprayed. So although it appears that skunks, according to the scientific literature, do fart, getting close enough to detect their flatulence is not recommended!

RED FOX

Scientific name (Species): Vulpes vulpes

DOES IT FART? YES

The red fox (*Vulpes vulpes*) has the greatest geographic range of any carnivore species, they are found across the entire Northern Hemisphere, from the Arctic Circle to as far south as Northern Africa. Red foxes have also been introduced to Australia, where they have sadly become one of the most problematic invasive species, eating many rare species of birds and mammals.

As members of the canid family, foxes regularly pass gas, and this only adds to a number of particularly smelly traits that foxes have. Foxes use a combination of scent glands and very strong-smelling urine to scent-mark their territories. Many foxes live alongside humans in urban areas and will often mark gardens and streets with their urine and faeces. Whilst this may be repulsive to humans, this is often particularly appealing to domestic dogs (page 74) who often enjoy rolling around in their scent, much to the disgust of their owners. In another blow to dog owners, foxes can have digestive parasites which upset their stomachs and cause explosive bouts of flatulence, and these can occasionally be transferred to the domestic dog, causing the same, incredibly smelly, symptoms.

FERRET

Scientific name (Species): Mustela putorius furo

DOES IT FART? YES

The ferret is a domesticated variety of the European polecat, *Mustela putorius*, a member of the *Mustelid* family native to Europe and Northern Africa. Ferrets were originally domesticated to hunt rabbits and small rodents, as their thin, elongated bodies are ideal for fitting into rabbit burrows and holes in the ground. They were brought over to the UK by the Romans in the first century but did not become popular as pets until the 1960s.

Appropriately for a species which has a scientific name that means 'angry smelly weasel', ferrets do fart, although that is not where they get the name. Ferrets have anal glands which produce a very strong odour – these are often removed in pets to prevent them smelling too strongly. This smell is only amplified by the pungent bouts of flatulence they regularly emit, especially during defecation and when stressed. Ferrets often surprise themselves with their own farts, and owners often report a confused look on their pet's face in the direction of their backside after they audibly pass gas. When ferrets are scared they will scream, puff themselves up, fart and defecate simultaneously. And to think some people put them down their trousers…

SEAL AND SEA LION

Scientific name (Clade): Pinnipedia

Does it fart? Yes

The clade *Pinnipedia*, more commonly known as pinnipeds, includes all species of seals, sea lions and walruses. There are 33 living species of pinniped in the world today, 15 species of eared seals and sea lions (*Otariidae*), 17 species of earless seals (*Phocidae*) and the walrus, which is the only living member of the *Odobenidae* family. These species feed on vast quantities of fish and, in a number of species, other marine invertebrates such as crabs. This is a surefire recipe for large quantities of potent, fishy-smelling farts (or in the case of leopard seals even penguin-smelling). Zookeepers have reported that sea lions in particular have the smelliest farts in the animal kingdom. When in and around seal and sea lion colonies the fishy smell can be incredibly strong, and this is at least in part to their bouts of flatulence, which are audible and often accompanied by numerous, equally fishy-smelling burps.

All species of pinniped live in aquatic habitats, and when observing these animals in the water they can often be seen passing gas. Researchers camping near seal colonies have even reported that the sounds of seal farts can keep them awake at night – something to which this author can certainly attest.

GUINEA PIG

Scientific name (Species): Cavia porcellus

DOES IT FART? YES

Guinea pigs were domesticated in approximately 5000 BC in South America; originally for eating, but upon being shipped to Europe in the sixteenth century they became a popular exotic pet species. Wild guinea pig relatives are known as cavies; the domestic guinea pig does not exist in the wild, and scientists are not certain which cavy species they descended from, although it is thought to be the montane guinea pig, *Cavii tschudii*.

Today guinea pigs are incredibly popular and are among the 10 most commonly owned pets in the UK. Anyone who has owned a guinea pig will know that they certainly fart, and these farts can be both noisy and smelly. Guinea-pig farts often make squeaking sounds, which can be difficult to distinguish from the vocalisations that these noisy, squeaky animals use to communicate. Broccoli, cauliflower and sugary foods, among others, cause guinea pigs to pass the most gas, and these small rodents often become more gassy with age. Too much gas in a guinea pig can be a bad sign, however, because if it becomes trapped it can cause serious health problems, so if your guinea pig is having trouble passing gas it is probably best to take it to the vet.

GRIZZLY BEAR

Species Name (Species): Ursus arctos

DOES IT FART? YES

'Does a bear poop in the woods?' runs the hypothetical question that many people are familiar with and, like the question 'Does it fart?', the answer for the grizzly bear is most assuredly yes, and the most probable location for these farts is also in woodlands. Grizzly bears, or the North American brown bears, are a subspecies of brown bears that ranges from Alaska through western Canada and into the north-western United States. Grizzlies, like all bears, are omnivores; while they will kill and eat other animals – typically fish and small mammals – they also scavenge and feed upon carrion, as well as plant matter and berries. Like their general diet, grizzly bears have unspecialised digestive systems, which means that hard-to-digest plant matter will appear in their faeces looking relatively similar to how it did after chewing.

While we are not aware of any research that has been conducted on the gas production and farting habits of grizzly bears, this large bear faces threats from humans' need for gas – namely natural gas – through the destruction of their habitat as well as direct mortality of bears when conflicting with workers to extract this non-renewable resource.

SEA CUCUMBER

Scientific name (Class): Holothurian

DOES IT FART? NO

There are currently 1717 holothurian, or sea cucumber, species recognised today. They are incredibly abundant; at depths of 9 kilometres and below they make up over 90 per cent of macrofauna (that's non-microscopic animals) biomass. Whilst sea cucumbers don't fart (they have very primitive digestive systems), there is plenty of fascinating stuff going on in and around sea-cucumber butts. Sea cucumbers breathe through something called a respiratory tree which is located in their cloaca (their equivalent to an anus).

Some species of sea cucumbers which inhabit coral reefs have developed an interesting, butt-related defence mechanism. When threatened by potential predators they eject sticky parts of their respiratory system, known as cuvarian tubes, through their rear end, in a process known as evisceration. These threads can entangle predators, meaning the sea cucumber can escape unharmed.

Although a sea cucumber's bum might seem an unattractive place to hang out, some fish would disagree. Some species of pearlfish (*Carapidae*) actually live inside the cloaca and respiratory trees of sea cucumbers, safe from predators. They get their food by slowly consuming the sea cucumbers' gonads when they get peckish. Although this is annoying for the sea cucumber, it isn't too damaging for them, as their incredible powers of regeneration mean their reproductive organs soon grow back.

ERROR
404
FART NOT
FOUND

BIRD

Species Name (Class): Aves

DOES IT FART? NO

The class Aves contains nearly 10,000 species of birds, which can be found on all seven continents and range in size from the ostrich (2.8 metres) to the bee hummingbird (5 centimetres), but none of them fart! Birds don't have the same gas-producing bacteria in their gut that are found in mammals and other farting animals, and food passes quickly through a bird's digestive system, which leaves no time for the build-up of toots. All the necessary anatomy is in place, though, so it is likely they could if they 'needed to'.

Although some people claim to have heard or maybe even have seen a bird fart (see parrot, page 18), there are good alternative explanations for these phenomena. So far the only scientifically documented record of a potential bird flatus comes from the thesis of a Cornell graduate student, Alan Richard Weisbrod, who, in much detail, recorded the behaviour of the blue jay (*Cyanocitta cristata*). He noted that on a cold day in December 1963, one of his study birds' defecation was accompanied by a 'small puff of whitish gas' which wafted 'below and parallel to the slightly raised tail' and then quickly dissipated; he also noted a second possible bird fart several days after his initial sighting. Unfortunately, this was likely just warmer water vapour from faeces meeting the cooler air and producing the visible gas.

LLAMA

Scientific name (Species): Lama glama

DOES IT FART? YES

Llamas are thought to have been domesticated around 4000 BC in the Andes mountains, and throughout the history of these Andean cultures they have been reared for their meat and used as pack animals. Today, however, they are much more widely kept and there are thought to be as many as 158,000 owned in the United States.

Llamas are well known for their spitting behaviour, which they use to assert dominance over others, however, a llama that has been raised properly will not spit at people. One less-famous activity carried out by llamas, though, is farting. This may be because this behaviour is not particularly frequent! Llamas are part of the *Camelidae* family, to which camels (page 88) also belong, and they have a similar digestive system. As a result they don't produce a whole lot of gas, and what they do produce is more commonly burped rather than farted out. Llamas with upset stomachs, on the other hand, are a different matter entirely, and they have been reported by llama owners to produce numerous bouts of flatulence. Luckily for the owners, llama farts, similar to llama poo – which is particularly dry, similar to rabbit (page 73) droppings – don't smell too bad.

SLOTH

Scientific name (Suborder): Folivora

DOES IT FART? NO

There are six species of sloth alive today, all of which live in the tropical forests of Central and South America. Sloths get their name because of their slow-moving lifestyle, however it is not just moving through the trees that they do slowly. Sloths have an incredibly slow digestive system, and it takes them days to process their food, which is entirely made up of leaves. This has an interesting side-effect; studies have recorded sloths defecating as infrequently as once every five days. This is probably a good thing, as sloths often climb down from the canopy in order to defecate, putting them at risk from predators.

Because of their leafy diet sloths have very simplified gut flora compared to other animals, which means that they don't produce flatulence. In fact, trapped gas is dangerous for sloths and a sign that something is wrong with their digestive system or diet. What their gut flora does do, however, is produce a whole lot of methane, but instead of being farted out it is absorbed through the gut and into the bloodstream before being breathed out. Sloths could potentially be the only mammal not to fart, however this may also be the case for other, less well-studied species and also bats (page 16), whose farting habits remain a mystery.

SALAMANDER

Species Name (Order): Caudata

DOES IT FART? MAYBE

Salamanders diverged from other amphibians approximately 200 million years ago and there are nearly 700 currently recognised salamander species. Despite this long history and wide distribution, to our knowledge no one has heard an audible salamander fart – but that doesn't mean we can't speculate.

Salamanders, like other amphibians (see frogs, page 69), may not possess strong-enough sphincter muscles to create the necessary pressure for a definitive flatus, but one species of aquatic salamander, the greater siren (*Siren lacertina*), has been found to have fermenting microbes within their intestines, which aid in the digestion of plant matter and may provide the necessary gaseous component for a salamander fart. Moreover, a common defence mechanism for salamanders is to defecate on their would-be attackers, and many scientists who study these amphibians are well aware of the surprisingly powerful stench that can accompany their faeces. In the eastern United States, some populations of terrestrial salamanders can be so abundant that their combined weight (biomass) exceeds that of all the small mammals and birds together in that same area, even though each salamander only weighs a couple of grams! So next time you go for a hike in the woods, you can be comforted by the fact that underneath your feet there might be hundreds of salamanders underground, quietly passing gas.

CHIMPANZEE

Species Name (Species): Pan troglodytes

DOES IT FART? YES

Chimpanzees are the closest living relative to humans, and in addition to sharing 98 per cent of our DNA chimps also share our ability to fart; in fact, chimpanzees fart loudly, often and without shame. When studying chimpanzees in their native habitat, within forested regions of western and central Africa, scientists have actually used the sound of these farts to locate individuals.

Just like humans, parasitic infections in chimpanzees can lead to gastrointestinal symptoms such as increased flatulence, but chimps are smart and know how to use tools. In the wild, chimpanzees have been observed to ingest liquid and fibrous material from the shoots of a tropical shrub called *Vernonia amygdalina* to alleviate gastrointestinal distress, as well as to reduce both flatulence and parasitic loads.

Researchers at Yale University have developed a cracker to provide captive chimps with their necessary nutritional needs, but this supplement was found to have a very interesting side-effect: the almost complete elimination of all chimpanzee farts. This occurs as a by-product of the cracker's high calories; captive chimps would eat a smaller amount of food and had to chew for longer, which reduces gas build-up in the intestines. Given their affinity for farts, we can only assume that the zookeepers responsible for caring for these chimps were quite pleased with this side-effect.

SILVER-SPOTTED SKIPPER

Species Name (Species): Epargyreus clarus

DOES IT FART? MAYBE

Within this book you will find examples of some impressive farting insects, such as termites (page 37), beaded lacewings (page 8) and American cockroaches (page 70), but when it comes to the silver-spotted skipper (specifically, the caterpillar form of this American butterfly) we are unsure if they can produce a fart. Nonetheless, while not technically a fart, these soon-to-be butterflies exhibit some impressive power in their hindquarters that is worth mentioning. The silver-spotted skipper exhibits high site fidelity (that is, it stays in one spot): it rolls a host plant around its body and remains in this small area during its whole larval period. This small living space, however, presents a problem, namely the build-up of waste. Fortunately, the skipper has evolved an impressive way to keep a tidy house. While small (only 4 centimetres in length), the silver-spotted skipper is able to build up blood pressure within its anal compartment to catapult its faeces (known as 'frass') up to 153 centimetres away from their bodies: equivalent to nearly 65 metres for a human! Research has shown that this behaviour has evolved to reduce predation; wasps are attracted to the skipper's frass odour and by ejecting these pellets a great distance they can hide from their would-be predators.

EASTERN HOGNOSE SNAKE

Species Name (Species): Heterodon platirhinos

DOES IT FART? PROBABLY

Like all snakes, the eastern hognose snake probably does fart. However, noxious odours are also used by these snakes to deter predation. When threatened, the eastern hognose snake will initially raise its head, puff out and flatten the skin around its neck and head, and hiss. If this aggressive behaviour fails, the snake takes an alternate approach by playing dead. Individuals will roll over, open their mouth and stick out their tongue, then emit a foul-smelling musk from their cloacal glands in the hopes of ruining their attacker's appetite. Though its displays are convincing, this non-venomous snake is all show, as it rarely bites, and if turned right side up while feigning death the snake will again roll onto its back. Playing possum is known for a few other species of snakes throughout the globe, although this behaviour is typically absent in captivity. The best displays, however, are those accompanied by extra smelly snake musk, which can persist in clothing even after multiple washes.

SAND TIGER SHARK

Scientific name (Species): Carcharias taurus

DOES IT FART? YES

The Sand tiger shark (*Carcharias taurus*) is known by a number of different names globally, including the grey nurse shark, blue nurse sand tiger shark and the spotted ragged tooth shark. As sharks are denser than water they sink to the bottom when they stop swimming. The sand tiger shark has found a unique solution to this problem. It has been observed in aquariums and in the wild gulping air from the surface of the water, and this air is then stored in the stomach, allowing the sand tiger shark to hang in the water column and maintain buoyancy. Reportedly sand tiger sharks release air through their cloaca when they want to become less buoyant, and a stream of bubbles can be observed. This is a similar mechanism to that used by herring (page 1) in communication, but a lot less noisy.

Sand tiger sharks have a formidable reputation, in part because of their fearsome, protruding teeth, but in reality these sharks are very docile and not aggressive, and therefore pose no threat to humans.

FROG

Species Name (Order): Anura

DOES IT FART? MAYBE

Frogs can make a variety of noises in the form of specific calls; those to attract mates, to ward off potential competing males, to send a distress signal if attacked and to warn others when startled. Frogs even have a release call; if a male tries to mate and engage in amplexus (the mating position where the male grabs the female around the back) with a female that has already released her eggs, or with another male, the uninterested frog will let this potential suitor know that their efforts to mate are in vain. However, one sound that frogs likely do not make is a fart. These amphibians do not possess very strong sphincter muscles, so any potential escaping gas through their cloaca may not cause the necessary vibrations against these muscles to produce an audible flatus. However, in the tadpole stage of some species researchers have found the presence of fermenting bacteria within their guts, which aid in the digestion of plant matter – the primary food source for these larval amphibians – and may produce intestinal gas.

Interestingly, scientists have observed that when using green tea leaves as a food source for captive reared tadpoles, individuals can get gas bubbles within their intestines, causing them to swim upside down for extended periods. Unless properly released, individuals can die from this gas attack, although for those tadpoles who do survive, the end from which this gas exits is unknown.

AMERICAN COCKROACH

Species Name (Species): Periplaneta americana

DOES IT FART? YES

Cockroaches have been around for about 280 million years, and during that time they have adapted to a wide variety of habitats throughout the world. They can survive temperatures as low as -122°C (-188°F), and they can live for up to a month without food and up to 45 minutes without air. Their heads can even survive for hours after they have been removed from their bodies! Unfortunately, one of their favourite places to live is wherever humans, and their food, happen to be. Cockroaches will eat anything they can find but prefer sweeter foods, and can decidedly ruin an entire food pantry. They reproduce rapidly; on average a female produces 15 eggs each month for 10 months, meaning it won't take very long for a colony to infest a home.

If this didn't put you off them enough it might help to know that the American cockroach farts, and has probably farted in your food. Additionally, these farts can contain methane. When controlling for size, larvae tend to produce more methane than the adults while, much like humans, high-fibre diets tend to lead to more gas production for these insects.

ORANGUTAN

Species Name (Genus): Pongo

DOES IT FART? YES

There are currently two species of orangutan recognised within the genus *Pongo*; unlike the other great apes, these species are restricted to Asia (Indonesia and Malaysia) and are primarily arboreal. Similar to other great apes, orangutans are closely related to humans and share around 97 per cent of our DNA as well as many similarities in physical appearance. In fact, the name 'orangutan' translates to 'person of the forest' and indigenous people often mistook orangutans for humans hiding amongst the trees.

Orangutans are also similar to other ape species in their ability to fart and their lack of shame in doing so. Orangutans are so fond of farts that they make noises from both ends, and have been observed making raspberry sounds, among many other different vocalisations. While the precise meaning for blowing raspberries is unknown, orangutans are often observed doing this during nest building before going to sleep. Copying this behaviour is not advisable for humans, however, as audible fart noises are typically not welcomed in bed.

RABBIT

Species Name (Genus): Oryctolagus

DOES IT FART? YES

Rabbits are described as non-ruminant herbivores, which means that while their diet consists of plant matter such as grass, flowers, as well as twigs, they do not have a specialised stomach (see cow, page 102, or goat page 2) to digest plant material. Instead, they rely on microorganisms (bacteria and protists) within their caecum (a pouch in their large intestine: you have one too) to extract nutrients from their cellulose-based diet. Since their food is initially digested in their large intestine, in order to get the most nutrition out of their food rabbits also reingest their cecotropes, which are soft faeces consisting of the fermented plant material.

Unsurprisingly, both the rabbit's slightly disgusting diet and their digestive system provide the perfect recipe for farts. Rabbits not only can and do fart, but they need to fart. Stress, dehydration and a diet that is low in fibre but high in carbohydrates and sugar can lead to a build-up of gas within their intestines, which is known as intestinal stasis. While farts are often humorous, this is no laughing matter for rabbits, as this gas build-up is extremely painful and can become fatal very quickly unless properly released, sometimes requiring medical intervention.

DOG

Species Name (Species): Canis lupus familiaris

DOES IT FART? YES

The phrase 'man's best friend' used in reference to the domestic dog is originally attributed to Frederick the Great of Prussia; however, it is doubtful that he was remarking on the dog's propensity to fart. All dog breeds fart, often with a rather pungent odour. However, not all canines are equal in their farting. For example, Boston Terriers, as a result of their shortened snouts, tend to swallow air more often than other dog breeds, which causes them to fart more often. Unfortunately, these terriers have a pleasant disposition and can be very affectionate, meaning they want to spend a lot of time around their humans, all the while passing odorous gas.

Given our close relationship with our canine companions, scientists have looked for new ways to reduce either the frequency or potency of their farts. Scientists have even developed a special coat that allows for the non-invasive collection of dog flatulence. The development of this product required a designated smell tester to rank each fart on how bad they smelled – so if you hate your current job, spare a thought for the fart tester. These pioneers in flatology found that diet supplements containing activated charcoal, *Yucca schidigera*, and zinc acetate reduced the intestinal production of hydrogen sulphide by up to 86 per cent, thus decreasing the frequency of malodorous farts!

PAINTED TURTLE

Species Name (Species): Chrysemys picta

DOES IT FART? YES

Currently there are nearly 300 recognised species of turtles, a group that is considered to be one of, if not the most endangered vertebrate taxa on this planet. This is due, in part, to habitat loss as well as collection for food and the pet trade.

The painted turtle is a widely distributed North American species of freshwater turtle – commonly referred to as 'terrapins' along with some brackish species – and like tortoises (page 86), this turtle species farts. Unlike tortoises, aquatic turtles face special problems concerning their farts: without their proper release, turtle buoyancy, much like the manatee (page 79) or Bolson pupfish (page 13), can be affected and individuals may not be able to dive. But for the painted turtle, as well as a few other aquatic turtle species, the cloaca isn't just for the release of gas, it also takes in gas, namely oxygen. The painted turtle accomplishes cloacal respiration through a specialised sac, known as a bursa, which can absorb oxygen whilst the turtle is underwater. Cloacal breathing is advantageous for aquatic turtles because it doesn't require their lungs, which, due to the turtle's shell, requires the use of many energy-costly muscles and, as a result, the increased production of lactic acid. This means that during hibernation painted turtles can burrow into mud under water and can remain there to avoid freezing temperatures.

COLOBUS MONKEY

Species Name (Genus): Colobus

DOES IT FART? YES

There are five species of colobus monkeys that can be found within forested regions of western and central Africa. All five species within this genus are herbivores and their diet consists of leaves, flowers and twigs, as well as unripe fruit. Interestingly, colobus digestive anatomy is similar to that of a cow's (page 102) and other hoofed mammals; these monkeys have relatively large, four-chambered stomachs where the first two sections function as a fermenting area. This unique diet and digestive anatomy results in three identifiable behaviours for these monkeys. First, they need to eat a lot as their leafy diet is typically low in nutritional value, which means they spend approximately 20–30 per cent of their day consuming food. Second, to save on energy colobus monkeys have long bouts of inactivity; adults will spend up to 60 per cent of their day sitting and resting. Thirdly, as you may have guessed, colobuses produce plenty of carbon dioxide and methane gas and, as with other primates, they have no shame in releasing this in the form of farts. Therefore, at any given moment during the day these lazy monkeys are likely either eating, resting or farting. Interestingly, a recent study suggests that sitting and resting also helps with expelling gas – though in the form of burps – due to colobuses unique digestive anatomy, as this resting posture prevents increased pressure on their respiratory organs and thoracic cavity!

WEST INDIAN MANATEE

Species Name (Species): Trichechus manatus

DOES IT FART? YES

If you thought an animal affectionately known as a 'sea cow' farted, you would be correct. The West Indian manatee does indeed fart – a lot. In fact, manatees make great use of their farts. First, manatees are herbivores, and their all-plant diet leads to plenty of gas production, especially methane. Second, the manatee's diaphragm, a major muscle involved in respiration, is quite different compared to other mammals; it consists of two hemidiaphragms, isn't directly connected to the sternum, is located dorsal to their hearts (closer to their back) and it extends horizontally throughout most of their body cavity. Additionally, the manatee's intestines have small 'pouches' throughout that allow for gas storage. This unusual anatomy allows the West Indian manatee to manipulate their farts as a mechanism for buoyancy. By storing gas within specific areas of their intestines, manatees can make their body more buoyant and float towards the surface, while the compression and release of flatulence makes these mammals sink. Farts are so crucial for manatees that constipated individuals cannot swim properly and have been observed floating with their tails higher than their heads.

SPOTTED HYENA

Species Name (Species): Crocuta crocuta

DOES IT FART? YES

Spotted hyenas are an extremely social species of mammal; they form large hierarchically structured groups and exhibit a wide variety of vocalisations to communicate. It is because of one of these vocalisations that the spotted hyena is commonly known as the 'laughing hyena'. While hyenas do in fact fart, this is not the reason for their 'laughter': this call is typically made when an individual is attacked or harassed by another hyena, generally over food.

The spotted hyena is primarily a hunter and is able to completely break down and digest bones, and because of this their faeces are white due to the high calcium content of their diet. While the typical prey for spotted hyenas are medium-sized hoofed mammals, they have been observed to eat smaller prey such as fish or birds and also go after large prey; one group has even been observed attacking and killing an adult hippopotamus (page 108). Anecdotal observations of hyena flatulence in the wild suggest that the worst-offending gas is produced after the spotted hyena eats camel intestines. Though the malodour of the hyena's farts, with respect to their diet, has not been rigorously studied, perhaps the contents of the camel's intestines provide the increased fuel as the hyena already eats a high-protein, all-meat diet.

BOBCAT

Species Name (Species): Lynx rufus

DOES IT FART? YES

The bobcat is a single species which has been subdivided into at least 12 subspecies, with much contention over these groupings. It has an expansive distribution, extending from southern Canada to southern Mexico with gaps within parts of the Midwestern United States. Throughout their range, bobcats are often mistaken for cougars (*Puma concolor*). These cats can readily be distinguished by a few characteristics; bobcats have pointed ears and are on average 2–3 times smaller than cougars. Bobcats are generalist carnivores as well as skilful hunters; their diet can consist of large mammals such as deer, but typically they prey upon smaller animals such as rabbits, birds and mice, as well as some reptiles.

This high-protein diet most assuredly leads to bobcat farts and of the smelly, sulphur-rich variety. Whilst their diet varies, anecdotally, a diet consisting primarily of squirrels appears to produce more pungent odours, though the reasons for this are uncertain at this time. Perhaps this phenomenon might be explained by squirrels containing a higher concentration of sulphur – another chemical compound that stimulates bobcat intestinal flora to produce more gas – but more research into this is clearly urgently needed.

PYTHON

Species Name (Family): Pythonidae

DOES IT FART? YES

In the wild pythons can be found in warm and wet habitats within Africa, Asia and Australia. However, due to many species' docile nature, unique colour morphs and ease of breeding, pythons can also be found in many people's homes as pets. Others, however, may not see the appeal. For one thing, pythons can get big; the Burmese python (*Python bivittatus*), which is a popular pet, can grow up to 6 metres in length and 180 kilograms in mass. The growth of these large non-venomous snakes sometimes give their owners a case of 'buyer's remorse' and are, unfortunately, sometimes released into non-native habitats such as the Florida Everglades. As such, populations of Burmese pythons have become established within the Everglades and are considered invasive and destructive; they have caused dramatic declines in many mammal species and have even been observed preying upon alligators.

As popular pets, observations of python farts are numerous and their flatulence has been described, charmingly, as 'thick and meaty'. Though this release of gas may go unnoticed initially, as they are often inaudible, the subsequent and powerful accompanying stench, due to the python's carnivorous diet, is readily detected.

CAT

Species Name (Species): Felis catus

DOES IT FART? YES

There has been some debate over whether cats are truly domesticated. For some animals, such as the pet dog (page 74), domestication is obvious. Dogs are incredibly tame and rely on their humans for care, and their domestication is also evident within their genome, which differs significantly from their wild ancestors. Cats, though, are considered 'semi-domesticated', as there is less divergence from wildcats and house cats and wildcats still interbreed. While house cats can be affectionate companions and often seem reliant upon their humans for food and care, they still continue to use their impressive hunting skills. In fact, research shows that free-ranging cats kill up to 3.7 billion birds and 20.7 billion mammals every year and have even caused the extinction of several species of birds, mammals and reptiles.

There is, however, no debate over whether cats fart. They do, and it often smells particularly strongly due to the fact that their carnivorous diet is rich in protein and contains a relatively high concentration of sulphur – and a by-product of this diet is sulphur-rich farts. However, fitting its semi-domesticated attitude, your cat probably doesn't care how bad you think its farts smell, and as a result won't make any effort to save you from their pungent aroma.

TORTOISE

Species Name (Family): Testudinidae

DOES IT FART? YES

Turtles in the family *Testudinidae* are more commonly known as tortoises. These turtles are marked by being completely terrestrial and are notably very slow-moving. In addition to their walking pace, tortoises do a lot of things slowly; some species, such as the Galapagos tortoise, can take up to 25 years to reach sexual maturity, and even their DNA can be slow. Evolutionary rates for turtles have been found to be slower than most mammals, and even other reptiles such as snakes.

On the other hand, in terms of farts tortoises are similar to other reptiles, as they can and do fart. Our knowledge of tortoise farts comes, in part, from some direct accounts; for example, female Mediterranean tortoises (*Testudo hermanni*) have been observed to fart immediately before laying eggs, while we also have anecdotal evidence of captive tortoise flatulence. Additionally, the tortoise diet and digestive system is conducive to producing gas; tortoises are mostly herbivores and have a hind-gut fermenting digestive tract similar to that of some mammalian herbivores (see horses, page 10; rhinos, page 24). Unfortunately, comparative studies of fart speed throughout the animal kingdom have not yet been conducted, so at this point we can only speculate that the tortoise may also be slower in its fart production than other groups of animals.

CAMEL

Species Name (Genus): Camelus

DOES IT FART? YES

There are currently three species of camel, two of which – the one-humped (C. *dromedarius*) and two-humped (C. *bactrianus*) – have been domesticated, while the wild Bactrian camel (C. *ferus*) can only be found in three small populations in its native range within the Gobi and Mongolian deserts. Camels are probably best known for their ability to thrive in extremely arid environments through some impressive adaptations.

Perhaps less well known is the camel's ability to produce methane-rich farts. Similar to ruminants (see cow, page 102), camels are herbivorous and use fermentation within their foregut to break down the cellulose in plant matter, however, their stomachs have only three chambers and they are more accurately classified as pseudoruminants. Because of this similarity in digestive anatomy, it was initially believed that camel methane production would be on a par with that of a cow. However, studies show that these pseudoruminants typically produce an overall lower amount of methane per kilogram of body mass. This difference can be explained by camels' decreased activity levels and lower food intake, even with unrestricted access to food and water; by eating less, camels only contribute 1–2 per cent of the methane produced by the much gassier cattle and other domestic ruminants in the same region. As with cows, most of this gas is released through the mouth, not the anus.

IGUANA

Species Name (Family): Iguanidae

DOES IT FART? YES

There are currently 42 recognised species within *Iguanidae*, a subfamily that includes iguanas *sensu stricto* (genus *Iguana*) as well as other related species, although classification of species within this family has been subject to much debate. Iguanas can be found throughout the North and South American tropics and subtropics, as well as the Galapagos, Antilles, Fiji and Tonga. The green iguana (*Iguana iguana*) is also an invasive species in some locations – non-native populations have been established within parts of the Caribbean as well as Hawaii, Florida and Texas.

Much like geckos (see page 90) and other lizards, iguanas fart. Rhinoceros iguana (*Cyclura cornuta*) farts have been described as 'wet'-sounding with an increased flatus frequency observed with higher fibre intake or with high loads of parasitic infection. Similarly, the black spiny-tailed iguana (*Ctenosaura similis*) has been observed to fart more often when fed a diet that is based more on plants than animal protein. Green iguanas, on the other hand, rarely consume animal protein in the wild and their diet is decidedly herbivorous, which typically leads to more farts. These iguanas are popular pets and, as such, there have been numerous observations of audible, and sometimes loud, farts, which are often observed during defecation.

GECKO

Species Name (Infraorder): Gekkota

DOES IT FART? YES

Geckos are the most species-rich group of lizards, at least 1,650 are currently recognised, which constitutes about 25 per cent of all lizard species. Geckos have some incredible adaptations that allow them to survive in a variety of habitats. Many gecko species have adhesive toe pads that are full of microscopic 'hairs', known as setae, which allow them to cling to any surface, including glass. These toe pads are so strong that the Tokay gecko (*Gekko gecko*) could hold over 450 times their body mass!

Though the scientific literature is scarce on gecko flatulence, as reptiles we can assume that geckos are able to fart. Moreover, because geckos are very popular pets, we have received reports confirming this assumption; for example, audible farts have been observed for crested geckos (*Correlophus ciliatus*), often preceding defecation. While gecko farts have been described as malodorous (that is, having a bad smell), it is difficult to separate these odours from their faeces, and we would require more research to confirm these anecdotes.

OCTOPUS

Species Name (Order): Octopoda

DOES IT FART? NO

Octopuses evolved at least 140 million years ago and currently comprise approximately one-third of all cephalopods – a class of animals that also includes squid, cuttlefish and nautiluses. Octopuses are unique marine invertebrates in that they display a great amount of intelligence. To the best of our knowledge though, octopuses don't fart. Although their digestion can be slow (12–30 hours depending on temperature and species), this lack of flatulence may be due to a lack of gas-producing gut flora. However, octopuses do feature an impressive mode of transportation that may seem similar to a fart: jet propulsion via their siphon. Octopuses can use their muscular body to expel water quickly through their siphon, a funnel-like structure that is also used to take in oxygenated water for respiration, to escape quickly from potential predators. This isn't the only 'pseudo-fart' in the octopus' arsenal though, as these cephalopods can release ink into the ocean, using its siphon to disperse it, which can confuse or potentially be toxic to predators and allow escape. In fact, octopuses use two different modes of inking: a dispersed 'smokescreen' ink that is meant to obscure its location from the predator, and a more condensed 'pseudomorph', which are smaller ink clouds mixed with a greater concentration of mucous, creating a glob with roughly the same appearance as the octopus for the predator to mistakenly attack.

MONGOOSE

Species Name (Family): Herpestidae

DOES IT FART? YES

The *Herpestidae* family, which are commonly referred to as mongooses (and not, sadly, as mongeese), currently contains 15 genera and 34 species, although some species, such as the meerkat (*Suricata suricatta*), aren't always commonly referred to as a mongoose. Though the mongoose, specifically the Indian Grey Mongoose (*Herpestes edwardsii*), is well known for its ability to kill venomous snakes such as cobras because of their agility and immunity to the venom, mongooses have a diverse and mostly carnivorous diet.

Mongooses most assuredly fart, and they have some powerful anal scent glands to boot; the stench can remain in clothing even after washing. Like the fossa (page 99), mongoose farts are the stuff of legends: Maasai beekeepers believe that the mongoose uses its farts to drive away bees so that it can raid their hives for the honey, while a Bedouin proverb, which concerns a dispute that creates an irreconcilable group, translates to 'the mongoose farted among them' – it was believed that a mongoose fart could disperse large herds of camel (page 88) and it would be extremely difficult to bring the herd back together.

GORILLA

Species Name (Genus): Gorilla

DOES IT FART? YES

There are currently two species of gorilla that are recognised: the eastern (*G. beringei*) and western (*G. gorilla*), both native to forested regions of subtropical and tropical Africa. Gorillas are the closest living relative to humans after chimpanzees (page 63) and bonobos, sharing at least 95 per cent of their DNA with us.

Zookeepers who have worked with gorilla husbandry are well aware of the powerful body odour that these great apes produce, and similar to the lemur (page 106), research shows that gorillas use their body odour for communication. More potent odours are emitted during more intense interactions, especially during threat displays. But body odour isn't the only smell emitted by these primates. Powered by their mostly herbivorous diet, with some insect supplements, gorilla farts can be very loud and, like other primates, are released without shame.

WOODLOUSE

Species Name (Suborder): Oniscidea

DOES IT FART? KINDA

Species of woodlice can be more commonly, and sometimes affectionately, known by many different names, including 'pillbug', 'roly-poly', 'doodle bug' or even 'cheeselog'. Typically, these nicknames apply to the common pillbug (*Armadillidium vulgare*), but there are well over 4,000 species of woodlice. These isopods are important for terrestrial ecosystems; they break down dead and decaying plant matter and through their slowly-decomposing faeces their presence leads to an overall sequestration of organic matter within soils. Woodlice also have an unusual way of eliminating their nitrogenous waste which, while technically not a fart, is impressive gas excretion nonetheless. In mammals, nitrogenous waste is converted into urea and excreted as a liquid, but woodlice excrete this waste as ammonia. By not converting ammonia into urea, woodlice are able to conserve more water and energy. Normally ammonia is toxic, but woodlice have a high resistance to it and can build up high ammonia concentrations within their tissues, where it is eventually excreted as a gas through their exoskeleton. Interestingly, studies have shown that woodlice primarily excrete this ammonia gas during the day, often in typically short bursts of a few minutes, but it can last as long as an hour or more – which is likely one of the longest-known 'farts' in the animal kingdom.

FOSSA

Species Name (Species): Cryptoprocta ferox

DOES IT FART? YES

If you have never heard of a fossa you are not alone, as even the scientists who study this elusive mammal find it difficult to detect within its native range, which encompasses most of the island of Madagascar. Its classification has been similarly problematic and it has been interpreted at different times as being either closely related to the mongoose (page 94) or to felines. The fossa is nonetheless a fascinating species which is exemplified by its Scientific name, *Cryptoprocta ferox*. Its Latin name, *Cryptoprocta*, means 'hidden anus', which refers to the fact that its anus is obscured by a scent-gland-containing anal pouch, while the species name *ferox* refers to its ferocity. This top predator hunts during both the day and night and eats a variety of mammal, bird and reptile species, although it appears to have a preference towards lemurs (page 106), which can make up 50 per cent of the fossa's diet.

Fossa flatus is much like the fossa: fierce. So much so that humans note that its pungent lingering odour can make one's eyes water. Indeed, Malagasy myths surrounding the fossa claim that its posterior perfume could wipe out an entire chicken coop!

SOFT-SHELL CLAM

Species Name (Species): Mya arenaria

DOES IT FART? NO

Soft-shell clams belong to the class of organisms known as *Bivalvia*, which are animals that are enclosed by a two-part shell that features a hinge, although these clams, as indicated by their namesake, have a thin and relatively easily broken shell. Soft-shell clams are probably best known in coastal towns of the north-eastern United States, where they frequently feature on restaurant menus. Though soft-shell clams do not fart themselves, shellfish allergies and intolerance, which are common in humans (page 118), can surely lead to some bad gas.

On the other hand, puking is well within this animal's repertoire. Similar to the octopus (page 93), these molluscs feature a siphon – technically a dual siphon – in which food-particle-containing water is drawn through their intake siphon. Food is then filtered by small hairs, known as cilia, and oxygen is extracted by their gills, then water is expelled by their exit siphon. When threatened, the clam will rapidly expel water and undigested food and retreat by digging deeper into the substrate. Researchers studying soft-shell clams are well aware of this behaviour, as they have often had 'clam puke' shot at them. This projectile vomiting is powerful enough to sometimes end up not only on, but under, researchers' clothing.

SNOW LEOPARD

Species Name (Species): Panthera uncia

DOES IT FART? PRESUMABLY

Snow leopards, as their common name suggests, are cold-adapted felines that can be found within mountainous regions of central and southern Asia. Compared to other felines, such as the lion (page 34) or cheetah (page 29), snow leopards have small rounded ears, thick fur and stocky bodies, which all help limit heat loss; deep nasal cavities, which allow them to warm respired air; and a thick tail that they can wrap around their bodies while they sleep to stay warm. Similar to other felines, the snow leopard is a skilful hunter and a carnivore; they use their immense paws and long tail to expertly navigate and balance on rocky cliffs to ambush their prey.

Partly due to their camouflage fur, these are elusive cats which are rarely seen or captured on film in the wild and their farts are no different – currently we have no direct confirmation of snow leopard flatulence. However, we can safely assume that the snow leopard, like other felines, farts, and we can further speculate that, due to their thick floofy fur, these farts are likely muffled.

COW

Species Name (Species): Bos taurus

DOES IT FART? YES

If any animal is famous for its farts, it's probably the humble cow. There are roughly 1.4 billion cattle throughout the world, two-thirds of which can be found in China, India and Brazil. As a ruminant, the cow's herbivorous diet is digested in stages within a four-chambered stomach. Food is initially chewed, swallowed and mixed within the rumen of the stomach, regurgitated – now known as cud – chewed and swallowed again, then it re-enters the stomach where the process of cellular digestion and fermentation via microbes begins. Through the digestion of this plant material a lot of greenhouse gases, notably carbon dioxide and especially methane, are produced and expelled by cattle – roughly 100–200kg of methane per cow per year! It is estimated that livestock, especially cattle, are responsible for approximately one-third of the greenhouse gas emissions produced by agricultural sources.

However, it's not all farting: although cows do indeed fart, these gases are generally released as burps, or expelled during respiration. Unsurprisingly, curbing cattle gas emissions has been a continuing field of research, involving diet alterations such as supplementing diets with a seaweed that can block methane production, or even transplanting gut microbes from potentially less gassy species such as the kangaroo (page 11).

DOLPHIN

Species Name (Infraorder): Cetacea

DOES IT FART? YES

The infraorder *Cetacea*, which includes dolphins along with whales (page 38) and porpoises, evolved during the Eocene (approximately 33–37 million years ago). Though these mammals are entirely aquatic, they are most closely related to the hippopotamus (page 108). Unlike hippos, however, dolphins are carnivores and their diet consists primarily of fish and squid. Dolphins are also cooperative feeders; groups of dolphins – known as pods – will surround and 'herd' schools of fish into a small area, or sometimes into shallow waters where individual dolphins can swim into and feed. Digestion in dolphins takes place within a multi-chambered stomach; the first stomach compartment (the 'forestomach') serves as storage and ingested food here can be regurgitated willingly, whereas the following compartments begin the process of digestion.

While dolphin farts may be silent, muffled by the surrounding water, bubbles emitting from their anus provide scientists with direct evidence of this tell-tale gas. Additionally, dolphin farts can often be smelly as a result of their diet (see also seals and sea lions, page 51). However, dolphins may not fart frequently; due to their high metabolic activity, food passes through their digestive tract very quickly, likely reducing the build-up of gas.

LEMUR

Species Name (Superfamily): Lemuroidea

DOES IT FART? YES

Lemurs are a diverse group of primates, with 101 currently recognised species, all of which are endemic to the island of Madagascar. Some species are very small – the adorably named Madame Berthe's mouse lemur (*Microcebus berthae*) is the world's smallest primate, reaching an average mass of only 30 grams – while the babakoto (*Indri indri*) weighs up to 9 kilograms. This diversity is also present in lemur ecology; although lemurs are all primarily arboreal, species can be either diurnal or nocturnal, and herbivores or omnivores.

Not surprisingly, lemurs do fart, but that isn't the only smell deployed by these primates. Lemurs use scents as a form of communication and many species have multiple scent glands located throughout their bodies. For example, the ring-tailed lemur (*Lemur catta*) has a wrist gland that produces a potent, but short-lived, clear odour, while a scent gland near their shoulder produces a longer-lasting brown odour that has the consistency of toothpaste. Male ring-tailed lemurs will use these glands to engage in 'stink fights', where they will rub the products of their two glands together and onto their tails, which they will wave above their heads, wafting their smelly concoction at their rivals and showing off their superior odour.

DIVING BEETLE

Species Name (Family): Dytiscidae

DOES IT FART? MAYBE

There are at least 4,000 species of diving beetles within the family. As their name suggests, diving beetles are aquatic and can be found within lakes, ponds and stream habitats. Both the larvae and adults are voracious predators, preying upon other invertebrates, such as mosquito larvae, as well as tadpoles and fish. They have also been observed preying upon animals larger than themselves. Diving beetles use their powerful hollow jaws to ensnare their prey and inject them with digestive enzymes, but their feeding modes vary with their life stages; larvae use their hollow jaws like a straw and suck the digested juices from the prey, whilst adults will tear off small chunks and swallow these pieces whole.

Based on available evidence, we are unsure if diving beetles fart, but they do feature some impressive gas-exchanging abilities at their rear end, reminiscent of the painted turtle (page 76), to allow them to remain submerged. Diving beetles need to breathe air, but they are able to trap and store oxygen in a bubble when they surface within a cavity underneath their wing casing, and are even able to replenish some of their oxygen supply whilst under water!

HIPPOPOTAMUS

Species Name (Species): Hippopotamus amphibius

DOES IT FART? YES

The common hippopotamus is one of two living hippopotamus species – the other is the pygmy hippopotamus (*Choeropsis liberiensis*) – and both species are native to Africa. Though the word hippopotamus means 'river horse', hippos are more closely related to dolphins (page 104) and whales (page 38). Hippos are probably best known for their immense size – males can be as large as 4,500 kilograms – and their aggression. Hippos are considered one of the deadliest mammals on the planet; if threatened they can attack using a combination of their huge body size, large canines and surprising speed (on land, hippos can run up to 30kph).

Hippos are primarily herbivores, though recent evidence suggests that they may eat meat more often than originally believed, including other hippos (though what this does to their farts is currently unknown). Like the camel (page 88), the hippopotamus is a pseudoruminant – it has a three-chambered stomach which lacks the rumen compartment and thus they do not regurgitate and re-ingest the cud – and they fart, often loudly. To mark its territory, the hippo will defecate while rapidly flailing its tail back and forth, which spreads its faeces. When defecation is accompanied by bouts of flatulence this behaviour can be especially amusing – or alarming – to onlookers.

KOALA

Species Name (Species): Phascolarctos cinereus

DOES IT FART? YES

The koala, much to the dismay of biologists, is sometimes referred to as a 'bear', although it is in fact a marsupial, a group of mammals that, among other distinguishing characteristics, feature a pouch to complete the development of their young, called joeys. Koalas are native to Australia and can be found farting within stands of gum trees (genus *Eucalyptus*) which serve as their food source – koalas are known to dine on around 30 of the 700-plus known species of gum trees, which may not be the best food source as these leaves are toxic to most animals. To break down toxic compounds produced by eucalyptus leaves koalas have specialised microbes within their digestive tract; these microbes are 'inherited' by joeys by consuming their mother's faeces. Eucalyptus leaves are also low in nutrients and, much like colobus monkeys (page 77), koalas rest a lot to save on energy – around 20 hours a day. As hind-gut fermenters, similar to horses (page 10), plant digestion takes place within their extra-long caecum of their large intestine – about 2 metres long – and food particles spend a lot of time here. To maximise the amount of nutrients extracted from their food, studies have found that koalas retain food within their digestive tract up to 100 hours in the wild and up to 200 hours in captivity – that's a lot of time in which to produce gas!

TAPIR

Species Name (Genus): Tapirus

DOES IT FART? YES

There are four living species of tapirs that can be found in forested regions of South and Central America as well as south-eastern Asia. A distinguishing characteristic of tapirs is their elongated snout, which is prehensile (capable of grasping objects) and which the tapir uses to hold branches and strip leaves or pick fruit, and can even function as a snorkel while underwater! Though somewhat superficially resembling a pig and often mistaken for an anteater or small hippopotamus (page 108), the tapir is more closely related to horses (page 10), zebras (page 31), rhinoceroses (page 24) and other ungulates that feature an odd number of toes. As such, tapirs are herbivores and hind-gut fermenters; in the wild they spend most of their time browsing vegetation across a variety of different habitats, often dispersing seeds of fruits across large distances.

Like their gassy evolutionary relatives, tapirs also fart and, anecdotally, scientists who have experience working with tapirs describe their flatulent habits as occurring in great amplitude (as we academics say).

MEGALODON

Species Name (Species): Carcharodon (Carcharocles) megalodon

DOES IT FART? NOT ANY MORE

Like the dinosaurs (page 32), the megalodon – a prehistoric giant shark – is extinct and has been for approximately 1.6 million years. When this monster was alive, though, it was truly formidable and is considered the largest shark to have ever lived. It reached up to 18 metres in length, with a jaw width of approximately 2 metres filled with five rows of teeth, each measuring around 18 centimetres long. Megalodon's bite force has been estimated to be greater than that of the *Tyrannosaurus rex*, approaching 182,000 Newtons (40,915 pounds force) – ten times that of the largest great white shark (*Carcharodon carcharias*).

As fossils cannot give us the answer to the question 'Does it fart?' we can look towards extant, or living, shark species, such as the sand tiger shark (page 66), for the megalodon and suppose that it potentially did fart, depending on its method of maintaining buoyancy. Due to its immense size, we can further speculate that if the megalodon did fart, these bouts of gas were also very large, and present, much like the megalodon, throughout most of the world's oceans during the middle Miocene (around 15.9 million years ago) to the late Pliocene.

WOMBAT

Species Name (Family): Vombatidae

DOES IT FART? YES

There are two genera of wombats that can be distinguished by their noses; the monospecific (single species within a genus) bare-nosed wombat (*Vombatus*) and the two species of hairy-nosed wombats (genus *Lasiorhinus*). Like the koala (page 111), the three species of wombats are herbivorous marsupials that are native to Australia. Unlike the koala, wombats are terrestrial, preferring to stay either within their burrows underground or foraging on the surface for food, typically at night. Wombats have a unique adaption to their fossorial lifestyle; unlike other marsupials the female's pouch faces backwards and opens up near its cloaca, which provides better protection for the young given the wombat's terrestrial lifestyle and prevents the pouch filling with dirt when digging. Unfortunately for the joeys (the marsupial young), this also means that they are in the position to provide confirmation that wombats fart. Though research is currently absent on wombat flatulence, their digestive anatomy is similar to the koala, in that they feature hind-gut fermentation and long food retention within their gut, so we can assume their farting behaviour is also likely similar and that being a baby wombat is an unpleasantly smelly experience.

WARTHOG

Species Name (Genus): Phacochoerus

DOES IT FART? YES

There are two species of warthogs: the desert (*P. aethiopicus*) and the more widely-distributed common (*P. africanus*), both of which are native to sub-Saharan Africa. Famously, warthogs have been portrayed in children's movies as pervasive farters, and while they do indeed fart, they are not quite the gassiest, nor the smelliest (see the sea lion; page 51) in the animal kingdom – not even close. Warthogs are primarily herbivores, however their diet may become more diverse and include insects or carrion (dead animals) if plant matter is scarce. Though an all-plant diet can be a good recipe for farts, it seems that the warthog's single-stage stomach and hind-gut fermentation, along with its high density and diversity of intestinal flora, create a very efficient digestive system for breaking down cellulose. In fact, one study found that an individual warthog produces as little as one-fiftieth of the methane produced by a giraffe (page 45), one-twenty-sixth of the elephant (page 27) and one-fifth of the zebra (page 31).

HAMSTER

Scientific name (Family): Cricetidae

DOES IT FART? YES

There are 26 species of hamster globally, living across Europe, Asia and the Middle East. Hamsters were not successfully bred in captivity until the 1930s, when zoologists found a mother and offspring of the Syrian hamster, *Mesocricetus auratus*, and brought them into a lab, where they rapidly reproduced (hamster gestation lasts only 18 days!). The Syrian or Golden hamster is the most popular of the five hamster species commonly kept as pets, however, despite millions of these animals being kept in captivity, there may only be under 2500 left in the wild.

Hamsters do fart, and it is well reported among pet owners that certain foods, such as cabbage, should be avoided due to the increase in gas production it causes. Gas can be bad for hamsters, which are prone to bloat. Food is important in hamster health and behaviour; they have to eat a variety of grains and vegetables to get a balanced diet. Anyone who has owned a hamster will know that they often store their food in cheek pouches in order to deposit it in a food store in their nest to be saved for leaner times. Syrian hamsters have particularly large cheek pouches which extend all the way to their hips. When full these pouches can double, or even triple, the size of a hamster's head.

HUMAN

Species Name (Species): Homo sapiens

DOES IT FART? YES

If you are reading this you are undoubtedly aware of the ability of the human species, like other primates, to fart. Unlike other primates, though, our species appears to experience emotions associated with our farts: namely shame, embarrassment, disgust, but also joy, schadenfreude (joy in other's misfortune) or perhaps even delight.

Humans have long been fascinated by their farts, which is exemplified by the odd myths and folklore surrounding them. For example, the water sprite, kappa, in Japanese folklore, is believed to be repelled by especially strong farts, while an Innu spirit of the anus named Matshishkapeu, which translates as 'the fart man', is a powerful and humorous spirit that is believed to be able to predict the future. Farts even appear in Dante Alighieri's famous poem *Inferno*, in which the demons' sign for the march was made by 'a trumpet of his ass'.

Not surprisingly, those humans who do not wish for their farts to be known have devised ways to blame others for their gas, and the dog (page 74) is a common scapegoat. However, all humans fart, and it happens every day, typically around 10 to 20 times each day, although this can be as high as 50 bouts of flatulence with increased flatus frequency, much like other animals, when diets are higher in fibre.

GLOSSARY

Abdomen
In vertebrates this is the part of the body containing the digestive system. In invertebrate arthropods it is the hind part, behind the thorax.

Allomone
A chemical substance produced by one species that affects the behaviour of a member of another species, to the benefit of the species that released the chemical.

Amplexus
The position in which frogs and toads mate – the male grabs onto the female's back and latches on.

Anaerobic
Something which happens in the absence of oxygen.

Anthropogenic
Caused by humans.

Archea
A type of single celled organism, similar to bacteria, but with a different cell structure.

Arthropod
Invertebrates with an exoskeleton, a segmented body and paired legs: includes insects, arachnids and crustaceans.

Autotomize
The ability of some animals to shed parts of their body to escape from predators before growing them back.

Bovid
Species in the family *Bovidae* – characterised by unbranched horns.

Bursa
A fluid-filled sac.

Ceacum
A pouch connected to the digestive system where the small and large intestines join. Can also be any body cavity that only has one opening.

Cellulose
A substance that is the main component in plant cell walls.

Cephalopod
The class of molluscs that includes octopuses and squid.

Cilia
Tiny, microscopic, hair-like structures.

Clade
A group of organisms that all descended from a common ancestor.

Cloaca
The end of the digestive tract in a large number of vertebrates that is used to excrete both faeces and urine, and sometimes also reproductive products (from the Latin for sewer).

Colon
The main part of the large intestine – this is where all the water and nutrients are absorbed.

Cud
Partly digested food that is brought back up to the mouth for further chewing in ruminants.

Diaphragm
The muscle that separates the chest cavity (the thorax) from the abdomen in mammals.

Dorsal
On the upper side of an organism.

Echolocation
Using sound to locate objects, used by animals such as bats, dolphins and some whales.

Endemic
Native and restricted to a particular area.

Fermentation
Chemical breakdown of a substance by bacteria.

Flora
In relation to 'gut flora': the symbiotic bacteria that live in the digestive system, particularly in the gut. Also known as intestinal flora.

Fore-gut
The part of the gut nearest the mouth.

Fossorial
Burrowing, or adapted for burrowing.

Frass
The faeces of insect larvae.

Gonads
The organ that produces sex-cells (sperm or eggs), generally the testes or ovaries.

Hemidiaphragm
Half of the diaphragm.

Hind-gut
The end of the gut nearest the anus or cloaca.

Invertebrate
An animal without a backbone.

Joey
Infant marsupial.

Lactic acid
An organic acid that is produced, among other ways, by muscles under anaerobic conditions. Acidic conditions caused by the natural production of lactic acid during strenuous exercise temporarily disrupts muscle function and prevents them from being overworked.

Macrofauna
Non-microscopic animals.

Marsupial

Animals within the infraclass Marsupialia native to Australia and the Americas. Though marsupials are mammals, they are distinguished from other mammals most notably by the presence of a pouch, which is used to complete their development.

Metabolism

The chemical and physical processes that sustains a living organism.

Methane

The main component of natural gas and considered one of the most potent greenhouse gasses. It has the chemical formula CH_4.

Miocene

An epoch of Earth's geologic history that lasted from approximately 23 to 5 million years ago.

Newtons

An unit of force: one Newton is the force needed to move one kilogram of mass at a rate of one metre per second^2.

Non-ruminant
Herbivores that do not have a four-chambered stomach and do not chew cud.

Oesophagus
The organ responsible for delivering food from an organism's mouth to its stomach.

Pliocene
An epoch of Earth's geologic history, from approximately 5 to 2.5 million years ago.

Primate
An order of mammals that includes monkeys, apes and humans. Primates are characterized by their large brains and grasping hands and feet.

Pseudoruminant
A group of animals that have similar fore-gut digestion to ruminants, but whose stomachs have three chambers as opposed to four found in ruminants.

Ruminant
A group of mammals that have a four-chambered stomach where food is fermented in the first chamber, regurgitated, and then re-ingested.

Setae
Hair- or bristle-like structures found on some organisms.

Siphon
Tube-like structures found in molluscs through which water or air flows.

Sphincter
A ring of muscles that closes an opening of the digestive tract, such as the anus.

Sputum
A mixture of saliva and mucus.

Stercoral sac
A sac at the end of a spider's digestive system, where moisture is extracted from food.

Subspecies
A taxonomic classification that is lower than a species. Subspecies are not divergent enough to warrant species status.

Taxon
A group of organisms that make up a unit of classification such as a species, family, or order.

Thorax
In vertebrates: the part of the body enclosed by the ribs between the neck and abdomen. For insects, the thorax refers to the area to which the legs and wings are attached.

Ungulate
Mammals with hooves.

Urea
An organic nitrogen-containing compound produced by an organism as a byproduct of metabolism.

Venomous
An animal that is capable of injecting another animal with a toxic substance (i.e., through a bite or sting).

Vertebrate
An animal with a spinal column.

Zooid
An animal that is a part of a colony.

Zoologist
Awesome people like Dani and Nick, who study animals.

Zoology
The study of animals and how they interact with their ecosystems.

ABOUT THE AUTHORS

DANI RABAIOTTI
@danirabaiotti

Dani is a zoologist currently studying the impact of climate change on African wild dogs (the best species of animal) at the Zoological Society of London and UCL. She is originally from Birmingham, UK, and is a proud Brummie. Dani has been mildly obsessed with animals and their behaviour since she was a small child, when her favourite animal was a crab and she would proudly proclaim she wanted to be a marine biologist. It was only more recently that animal farts came onto her radar, but, never one to pass up an opportunity to find out more about science and animals, she enthusiastically got stuck into researching and writing *Does it Fart?* Upon telling her family about the book her dad was mostly just excited that her co-author also has an Italian surname.

NICK CARUSO

@plethodonick

Nick is an ecologist currently in the Department of Biological Sciences at the University of Alabama, where he studies the role of climate in population biology of Appalachian salamanders. He is originally from St. Charles, Missouri, and grew up catching all manner of reptiles and amphibians while playing in forested areas, creeks and streams with his brother. While he has never studied animal farts or thought it would be a part of his life, Nick has always found humour in flatulence and, like Dani, could not pass up an opportunity to learn and write about farts. While researching the various animals for this book, he has found a new appreciation for farts and for the fascinating adaptations throughout the animal kingdom.

ETHAN KOCAK
@blackmudpuppy

Ethan Kocak is an artist and illustrator best known for the online graphic novel series *Black Mudpuppy* and various science-related sequential art projects. His works have been featured on the websites for Scientific American and WIRED Science, and tend to focus on reptiles and amphibians. He lives in Syracuse, New York with his wife, son and collection of rare salamanders.

ACKNOWLEDGEMENTS AND CONTRIBUTORS

Huge thanks to Science Twitter in general and in particular to the following animal experts for their contributions to this crucial science project:

Adriana Lowe (@adriana_lowe) – Aditya Gangadharan (@AdityaGangad) – Alex Bond (@TheLabAndField) – Alex Evans (@alexevans91) – Amy Schwartz (@LizardSchwartz) – Angie Marcias (@HereBeSpiders11) – Anthony Caravaggi (@thonoir) – Arjun Dheer (@ArjDheer) – Becky Cliffe (@BeckyCliffe) – Brian Wolven (@BrianWolven) – Carina Gsottbauer (@CarinaDSLR) – Cassandra Raby (@CassieRaby) – Chris Conrod (@edosartum) – Chris Pellecchia (@SquamataSci) – Dave Hemprich-Benett (@hammerheadbat) – David Steen (@AlongsideWild) – Ellen Holding (@pakachusus) – Erin Kane (@Diana_monkey) – Gregor Kalinkat (@gkalinkat) – Helen O'Neill (@hmkoneill) – Helen Plylar (@SssnakeySci) – Imogene Cancellare (@boliogistimo) – Ivan Daum (@ivandaum) – Jeff Clements (@biolumiJEFFence) – Jenny Gumm (@jennygumm) – John Smutko (@Smutt235) – Julie Blommaert (@Julie_B92) – Julie Wright (@indik) – Julien Fattebert (@FattebertJ) – Kim Kennedy – Lauren Robinson (@Laurenmrobin) – Lewis Bartlett (@BeesandBaking) – Lea Mac (@tecklen) – Mark Scherz (@MarkScherz) – MichaelReid (@mjcreid) – Nadine Gabriel (@NadWGab) – Natick Bobcat (@NatickBobCat) – Noah Mueller (@nbystoma) – Rachel Hale – (@_glitterworm) – Sarah McAnaulty (@SarahMackAttack) – Sergio Henriques (@SS_Henriques) – Sloth Sanctuary (@SlothSanctuary)

First published in Great Britain in 2017 by Quercus.

Quercus Editions Ltd
Carmelite House
50 Victoria Embankment
London EC4Y 0DZ

An Hachette UK company

A CIP catalogue record for this book is available from the British Library

HB ISBN 978 1 78648 826 8
Ebook ISBN 978 1 78648 827 5

10 9 8 7

Design by Rich Carr, Carr Design Studio

Printed and bound in the UK by Clays Ltd, St Ives plc